MATH Challenge 2

Published by InSTEP Education Publishing, Inc.
PO Box 1912 Coppell, TX 75019

InSTEP Education Publishing, Inc., Attention: Permissions Department

ISBN 978-0-9905149-6-1

For information about custom editions, special sales, premium and corporate purchases, please contact InSTEP Education Publishing at (214) 206-5926 or customerservice@instep.education.

MATH Challenge 2

The exercises within this workbook were created by highly-qualified educators and are designed to provide students practice with a variety of math skills, including fact recognition, basic operations (addition, subtraction, multiplication, and division), pattern recognition, and logical analysis.

The goal of each exercise is to encourage students to think critically to determine patterns and visualize mathematics outside of the standard format. The exercises are divided into three sections based on challenge level, ranging from easy to complex, and increase in difficulty to provide exposure, development, and practice with logical application of mathematics. Encourage your child or student to attempt as many exercises as possible to enhance skills.

Students are also provided the opportunity to apply heavy use of cognitive skills as they create their own puzzles to share with others.

The Math Challenge workbook is great for daily warm-up, enrichment exercises, or regular skill maintenance. We ask the question, *How many puzzles can you solve?*

Created & Produced by Teachers

InSTEP Education Publishing Inc.

www.InSTEP.Education

Table of Contents

Challenge Level I   [Easy] - pg 4

Level I exercises include many basic facts, simple operations, and patterns to solve. Students practice using learned skills to produce solutions.

Challenge Level II   [Medium] - pg 16

The Level II exercises are slightly more difficult than Level I, requiring students to apply fact knowledge and problem-solving skills to solve correctly.

Challenge Level III   [Complex] - pg 26

Exercises within Level III are more complex as they stimulate critical thinking skills and require students to sharpen problem-solving abilities to provide accurate solutions.

Create Your Own Puzzles - pg 33

Students create their own puzzles to share and challenge friends!

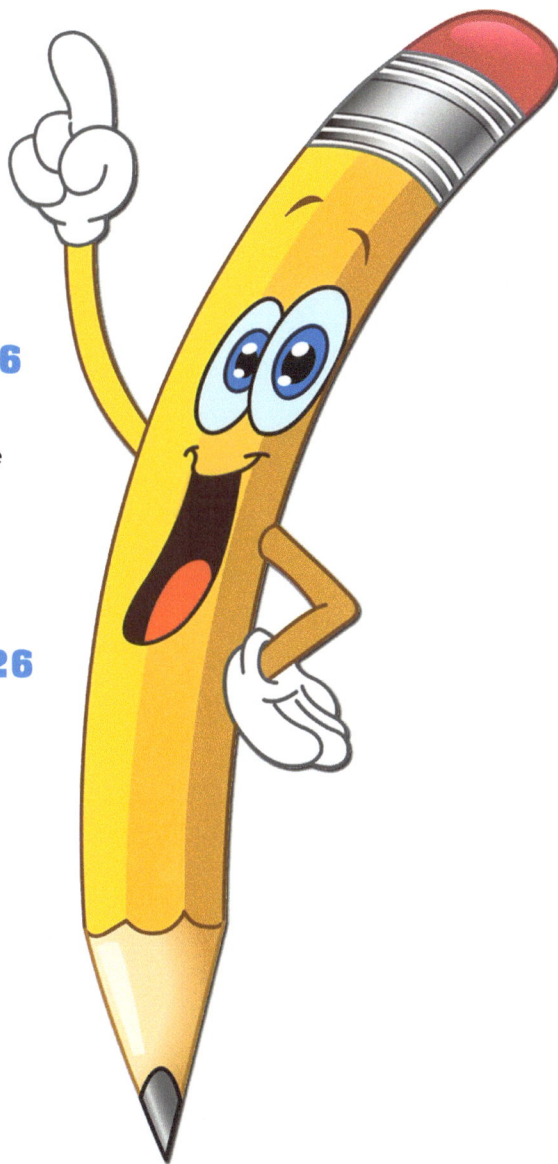

Popcorn Patterns

Directions: Can you solve the popcorn pattern puzzles below? Write the missing numbers on the blanks below.

1.

10

5   13

22

11   25

6   ___

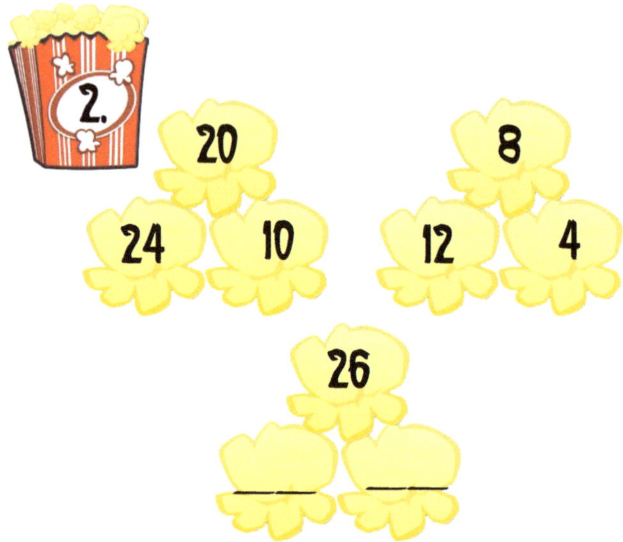

___

2.

20

24   10

8

12   4

26

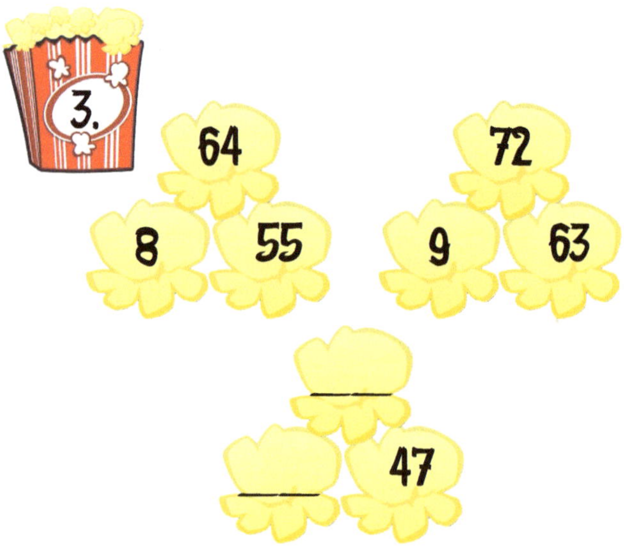

___   ___

3.

64

8   55

72

9   63

___

___   47

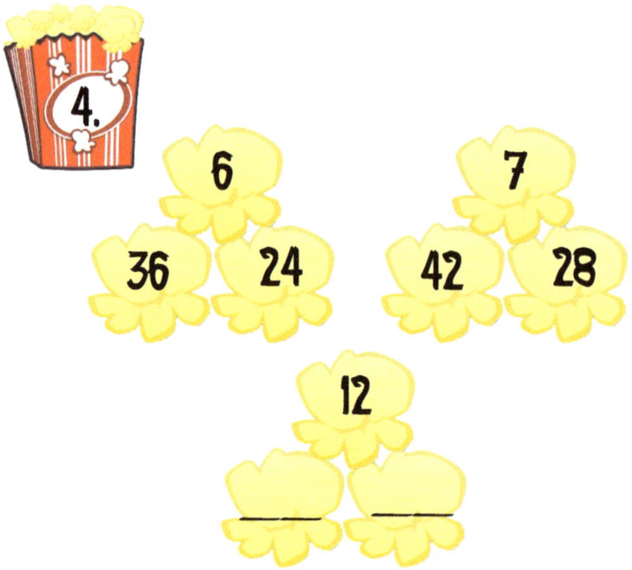

4.

6

36   24

7

42   28

12

___   ___

4

Pattern Word Problems

Directions: Read each word problem. Determine the pattern to solve.

1. Dasia likes to find and collect coins around her house. She collected 2 coins on the first day, 3 coins on the second day, 6 coins on the third day, 11 on the fourth, and 18 on the fifth day. If this pattern continues, how many coins will Dasia have collected by Day 7?

Day 1	Day 2	Day 3	Day 4	Day 5	Day 6	Day 7
2	3					

2. Austin enjoys playing video games. He started keeping track of how many points he scored on each level of his favorite game. Using the record below, how many points can Austin expect to score on Level 5?

Level 1	Level 2	Level 3	Level 4	Level 5
5	15	45	135	?

3. Gloria bakes cupcakes to sell after school. She baked 80 cupcakes in April, 88 in May, and 96 in June. How many cupcakes did Gloria bake in the months of July and August?

April	May	June	July	August
80	88	96	___	___

4. Sanjay's after-school club held a fundraiser selling popcorn for eight weeks. Sanjay is in charge of keeping the sales records. Can you fill in the missing information?

Week 1	Week 2	Week 3	Week 4	Week 5	Week 6	Week 7	Week 8
	34	46			82	94	

Place Value Puzzle

Directions: Use the clues and number bank below to complete the puzzle.

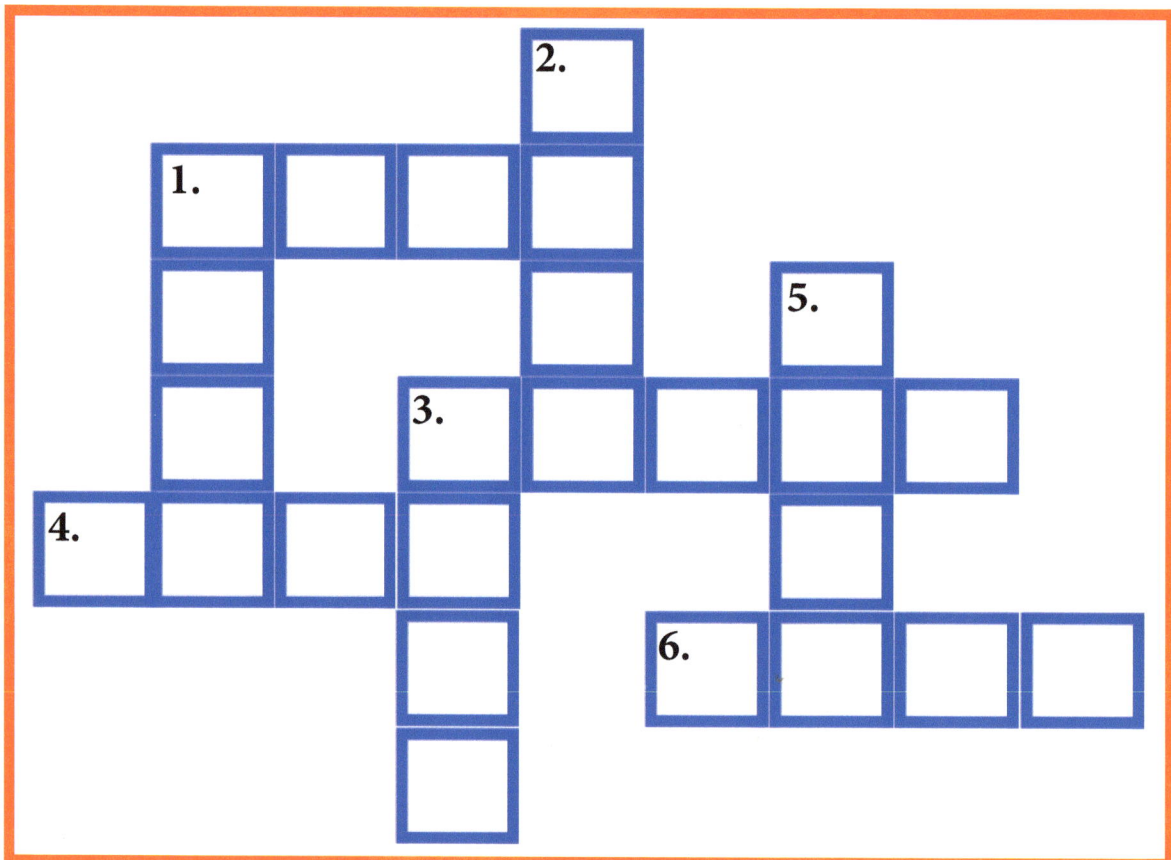

Across:

1. I have no hundreds

3. I have a digit in the ten-thousands place

4. My thousands place is larger than 8

6. I have an even digit in the hundreds place and an odd digit in the tens place.

Number Bank

9, 173	3, 749
2, 087	19, 156
1, 357	3, 502
2, 641	8, 293

Down:

1. I have a 1 in the ones place

2. I am an odd number with a 3 in the thousands place

3. I have a 5 in the tens place

5. My tens and ones places are both even numbers

Missing Digits

Directions: Complete each addition problem below by finding the missing digits.

1.
$$
\begin{array}{r}
8\ 6\ 7 \\
+\ 1\ \square\ \square \\
\hline
9\ 8\ 8
\end{array}
$$

2.
$$
\begin{array}{r}
\square\ 1\ \square \\
+\ 2\ \square\ 3 \\
\hline
8\ 9\ 5
\end{array}
$$

3.
$$
\begin{array}{r}
\square\ 2\ 8 \\
+\ 3\ \square\ \square \\
\hline
4\ 7\ 7
\end{array}
$$

4.
$$
\begin{array}{r}
6\ \square\ \square \\
+\ \square\ 7\ 6 \\
\hline
8\ 8\ 0
\end{array}
$$

5.
$$
\begin{array}{r}
\square\ \square\ 5 \\
+\ 4\ \square\ \square \\
\hline
5\ 5\ 5
\end{array}
$$

6.
$$
\begin{array}{r}
5\ \square\ 1 \\
+\ \square\ 2\ \square \\
\hline
9\ 0\ 0
\end{array}
$$

Hexagon Pyramids

Directions: Can you find the missing numbers? Find the sum of each top block by adding the two blocks directly beneath it.

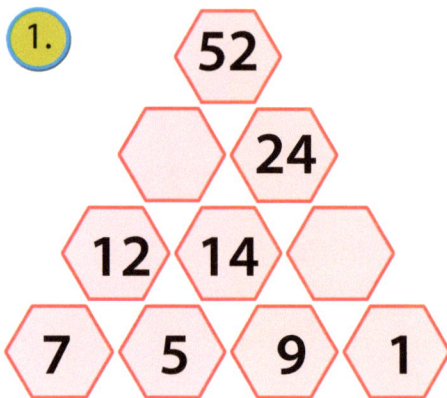

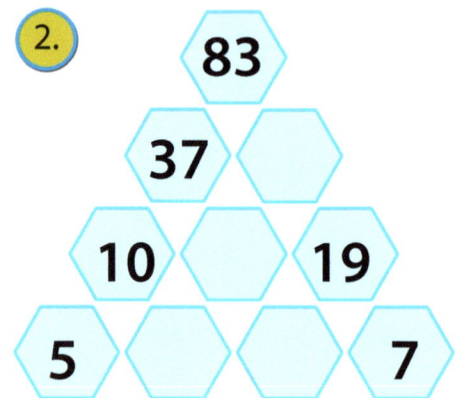

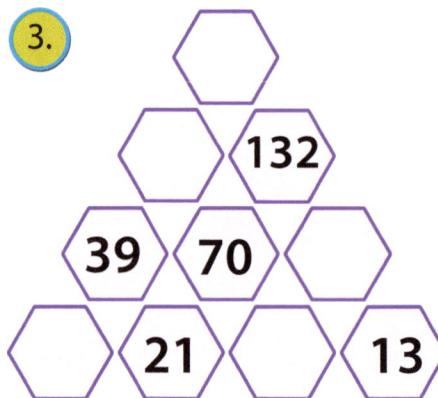

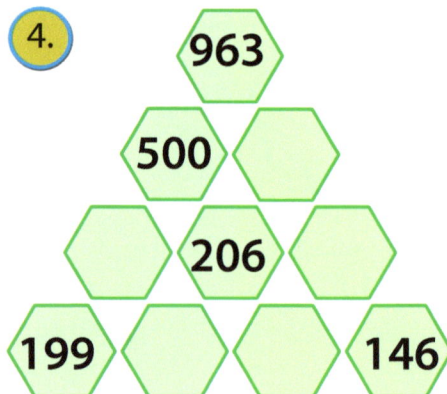

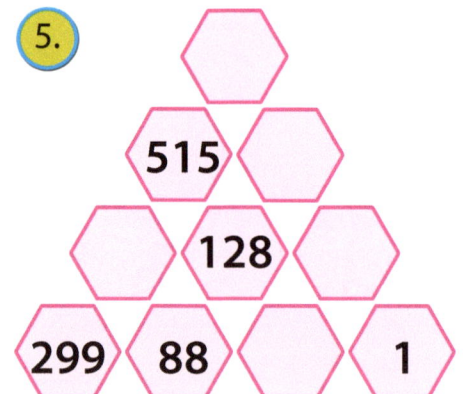

1.

52

24

12 14

7 5 9 1

2.

83

37

10 19

5 7

3.

132

39 70

21 13

4.

963

500

206

199 146

5.

515

128

299 88 1

8

Directions: Use the cards below to create five different number sentences that equal to 20. You must use only four cards for each number sentence.

8 2 5 7 4 9

EX: 8 ÷ 2 + 9 + 7 = 20

1. ☐ ☐ ☐ ☐ ☐ ☐ ☐ = 20

2. ☐ ☐ ☐ ☐ ☐ ☐ ☐ = 20

3. ☐ ☐ ☐ ☐ ☐ ☐ ☐ = 20

4. ☐ ☐ ☐ ☐ ☐ ☐ ☐ = 20

5. ☐ ☐ ☐ ☐ ☐ ☐ ☐ = 20

Directions: Complete each pattern by writing the correct numbers in the boxes or circles below.

1.
| 11 |
| 19 |
| 18 |
| 26 |
| 25 |
| 33 |
| |
( )

2.
| 6 |
| 18 |
| 20 |
| 60 |
| 62 |
| |
| |
( )

3.
| |
| 89 |
| |
| |
| 89 |
| 79 |
| 84 |
(74)

4.
| 1 |
| 25 |
| 12 |
| 36 |
| 23 |
| |
( )

5.
| 1 |
| 1 |
| 2 |
| 6 |
| 6 |
| 12 |
| 36 |
( )

Symbol Solve

Directions: Fill in the missing numbers below.

1.

$\square + \square + \square + \square = 28$

$\triangle + \triangle + \square = 11$

$\heartsuit + \square + \square + \triangle = 16$

$\heartsuit = \underline{\quad}$

2.

$\triangle + \triangle + \bigcirc = 28$

$\bigcirc + \bigcirc + \bigcirc + \bigcirc = 32$

$\triangle \times \bigcirc + \square = 84$

$\square = \underline{\quad}$

$\triangle = \underline{\quad}$

$\bigcirc = \underline{\quad}$

$\square = \underline{\quad}$

3.

$\square \, \bigcirc = 17$

$\triangle - \bigcirc = 4$

$\triangle + \triangle + \triangle = 15$

4.

$\triangle = 5 \times \heartsuit$

$\square\,\square\,\square\,\square = 40$

$\square + 20 = \triangle$

$\triangle + \square + \hexagon + \heartsuit = 50$

$\heartsuit = \underline{\quad}$

$\hexagon = \underline{\quad}$

$\square = \underline{\quad}$

$\triangle = \underline{\quad}$

11

Building Numbers

Directions: Use the digits in the box below to build the numbers described in each clue.

1. Build the smallest 3 digit number: _____

2. Build the smallest even number: _____

3. Build the largest odd number: _____

4. Build the largest 4-digit number with an even number in the hundreds place:

   _____

5. Build the largest odd number with an even number in the ten-thousands place:

   _____

Math Challenge

Mystery Number

Directions: Use the clues to find the mystery number.

1. If you first add me to 15, then divide by 5, your final answer will be 5.

Who Am I?

_____

2. First add me and 7. Then subtract 10 and divide by 2. Your final answer will be 9.

Who Am I?

_____

3. First subtract 20, then divide by 2. Subtract 10, and the final answer will be 30.

Who Am I?

_____

4. Multiply me with 12, then subtract 4, and divide by 8. The final answer is 7.

Who Am I?

_____

5. If you first divide me by 3, then add 91, and divide by 11, the final answer will be 11.

Who Am I?

_____

MATH ★ Challenge 2

Congratulations!!

_____
(Name)

has completed and mastered the
Math Challenge Level I!

_____
(Signature)

_____
(Date)

Price Tag Riddles

Directions: Arnez and Sasha are looking at the prices of different toys in the store. Use the numbers given and the clues to determine the correct price that belongs on each tag. Write the answer on each price tag. *Note: Must use dollars and cents.

1. 4, 8, 2
The highest price that can be made.

$.

2. 3, 7, 6
The price is the greatest odd number that can be made.

$.

3. 9, 7, 5, 2
The price is the lowest even number greater than $40.

$.

4. 1, 3, 4, 7, 8
The greatest number possible. The first and last digits are odd. Price is greater than $200, but less than $600.

$.

5. 2, 5, 1, 0, 4, 6
The greatest even number. The price is more than $500, but less than $1200.

$.

Cellular Patterns

Directions: Read each word problem. Determine the pattern used and solve.

1. Jayde wants to reduce the amount of minutes she uses talking on her cellphone every month. Using Jayde's notes, how many minutes will she use in the months of January and February combined?

My Minute Usage	
September:	315
October:	215
November:	135
December:	75

2. Divya usually keeps track of her monthly cellphone bill, but she missed recording a few months. Can you help fill in the missing months?

My Bills
April: $42.50
May: --
June: $65
July: $76.25
Aug: $87.50
Sept: --

3. Ethan and Chad are brothers who share a data package on their parents' account. Their shared limit is 100 text messages a month. Can you fill in the missing information? Which months did the boys go over their limit?

Text Messages		
Month	Ethan	Chad
1	60	20
2	20	40
3	36	12
4	15	30
5	33	11
6	50	100
7		25
8	33	
9	90	
10		70

4. Fernando began working in a cell phone store, selling smart phones. Using the record of his sales, can you determine how many more smart phones Fernando sold in December than in May?

Fernando's Sales
June 9
July 19
August 39
September 79
October 159

Missing Digits 2

Directions: Complete each subtraction problem below by finding the missing digits.

1.

```
    7   9   [ ]
  -     [ ]   6
  ─────────────
    7   6   7
```

2.

```
      [ ]   4
  -   5   [ ]
  ───────────
      2   5
```

3.

```
    3   [ ]   [ ]   9
  -       2    0   [ ]
  ───────────────────
    3   2    1    1
```

4.

```
    9   [ ]
  -  [ ]  7
  ─────────
    7   9
```

5.

```
    [ ]   0   [ ]
  -      [ ]   8
  ───────────────
    5    5    4
```

Rolling Dice

Directions: Four students rolled a pair of dice for a project in class. Look at what each person rolled. Can you predict what their next roll will be? Write each answer in the boxes below.

1.

2.

3.

4.

Directions: Use the order of operations (PEMDAS) and each set of cards below to write an equation that equals 25. The numbers in each equation must be used in the same order as they appear.

1.

_____ = 25

2.

_____ = 25

3.

_____ = 25

4.

_____ = 25

5.

_____ = 25

6.

_____ = 25

Note: There are no exponents used in the equations above.

Directions: Fill in the missing numbers below.

1.

$$\text{☆} + \text{◻} + \text{◻} + \text{◻} = \text{⬡}$$

$$\text{▲} \div \text{⬡} = \text{O}$$

$$6\,\text{☆} = 18$$

$$\text{◻}\,\text{◻}\,\text{◻} = \text{▲} + \text{◻}$$

☆ = _____

⬡ = _15_

▲ = _____

◻ = _____

O = _____

2.

$$\text{◻}\,\text{●} = 16$$

$$4\,\text{◻} = \text{▲} + \text{●}$$

$$\text{▲} \div \text{●} = 15$$

▲ = _____

● = _____

◻ = _____

3.

$$32 \div \text{▲} = 3 + 1$$

$$\text{◻} - \text{⬡} = \text{▲} \times 12$$

$$\text{⬡} + \text{⬡} + \text{⬡} = 72$$

$$\text{◻} - (\text{▲} + \text{⬡} + \text{♥}) = 40$$

♥ = _____

⬡ = _____

◻ = _____

▲ = _____

MATH ★ Challenge

Who Am I Riddles

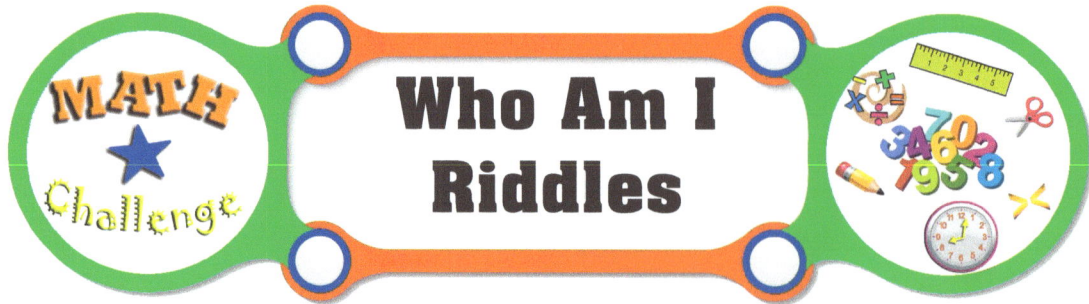

Directions: Can you determine the number each riddle is describing? Write the correct answer on the lines provided.

1. I am between 10 and 40. My two digits are odd. The sum of my digits is greater than 10.

_____

2. Two of my digits are even, but my tens place is an odd digit. The product of my digits is 96. I am between 425 & 455.

_____

3. I am less than 300 and all my digits are odd. The product of my digits is between 10 & 20. What two numbers could I be?

_____

MATH Challenge 2

Congratulations!!

_____
(Name)

has completed and mastered the
Math Challenge Level II!

_____
(Date)

_____
(Signature)

PEMDAS Sentences

Directions: Use PEMDAS rules to create equations below. Write parenthesis, +, -, ÷, or = in or around the boxes to make each number sentence true. You may use operations more than once, if necessary.

1. 8 ◯ 3 ◯ 1 ◯ 23

2. 8 ◯ 5 ◯ 2 ◯ 6

3. 4 ◯ 5 ◯ 3 ◯ 2

4. 9 ◯ 3 ◯ 5 ◯ 8

5. 16 ◯ 8 ◯ 2 ◯ 1

6. 12 ◯ 4 ◯ 48 ◯ 1

7. 12 ◯ 3 ◯ 2 ◯ 2

8. 14 ◯ 2 ◯ 5 ◯ 2

9. 5 ◯ 3 ◯ 10 ◯ 2

10. 6 ◯ 8 ◯ 2 ◯ 2

24

Note: There are no exponents used in the equations above.

Shapes & Sums

Directions: Complete the chart below by filling in each row and column with the missing sums.

♥	★	△	☀	36
♥	♥	♥	☀	42
★	O	O	☀	
★	△	O	☀	
♥	♥	★	☀	35
		50	30	178

☀ = ____   O = ____   ♥ = ____

★ = ____   △ = ____

MATH ★ Challenge

Complete the Chart

Directions: Fill in the missing shapes and numbers in the chart below.

♥			▲	
▲			★	25
♥	♥	♥	♥	16
▲	☀	▲	★	34
♥		☀	★	30
24	31	35	37	

Jot down the value of each shape:

♥ = ________ ★ = ________

▲ = ________ ☀ = ________

Rolling Dice 2

Directions: Math students rolled a pair of dice for a class project. Look at the results. Can you predict what the next roll will be? Write each answer in the boxes below.

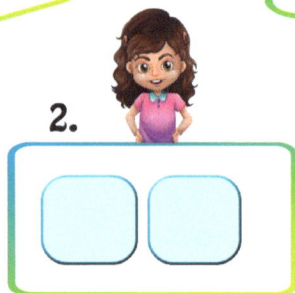

1. [2][6] , [4][5] , [1][4]

2. [5][1] , [4][2] , [3][3] , [2][4]

3. [2][3] , [6][4] , [3][5] , [6][4]

4. [6][6] , [4][5] , [5][4] , [3][3] , [2][4]

5. [4][1] , [1][4] , [5][2] , [3][5] , [4][3] , [2][6]

1.
2.
3.
4.
5.

Ladder Climb

Directions: Start from the bottom, and work your way up! Use the given numbers to create an equation that equals 75. You must use all operations (+, -, x, ÷). *Use each number and operation only once in each equation!

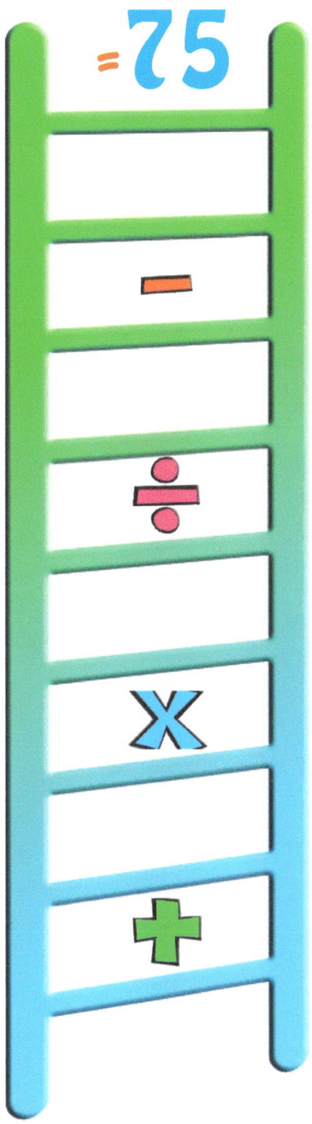

= 75

−

÷

X

✚

1. 0, 2, 7, 8, 10

= 75

2. 4, 4, 5, 5, 80

= 75

3. 2, 4, 5, 30, 50

Wave Patterns

Directions: Complete each wave pattern problem.

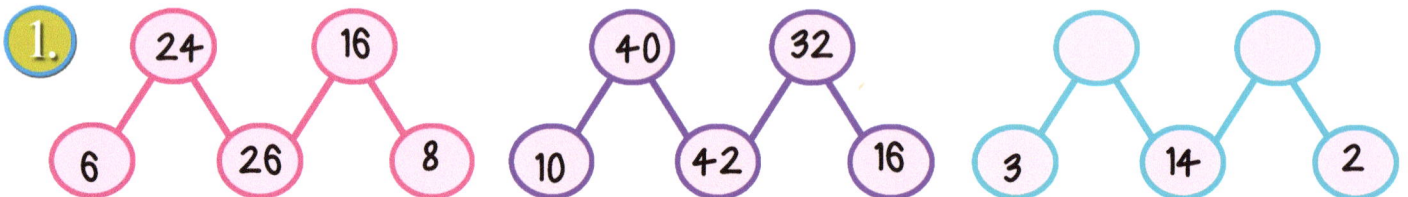

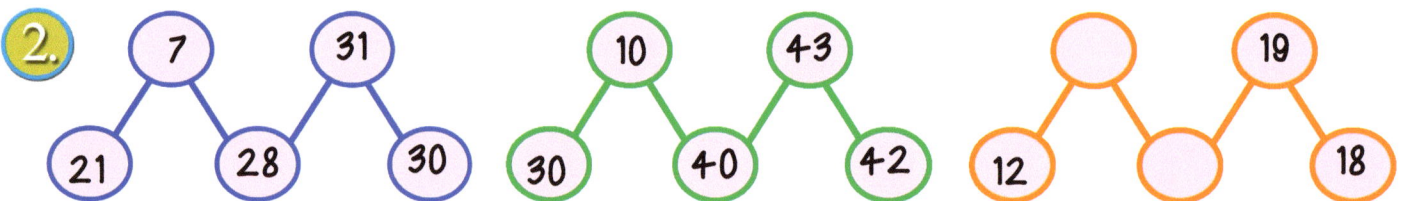

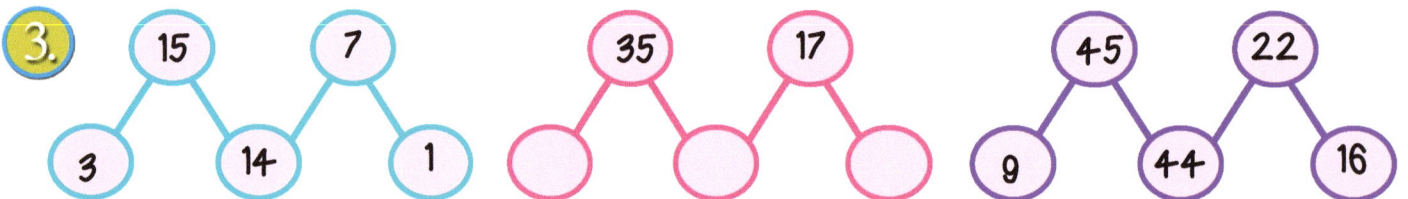

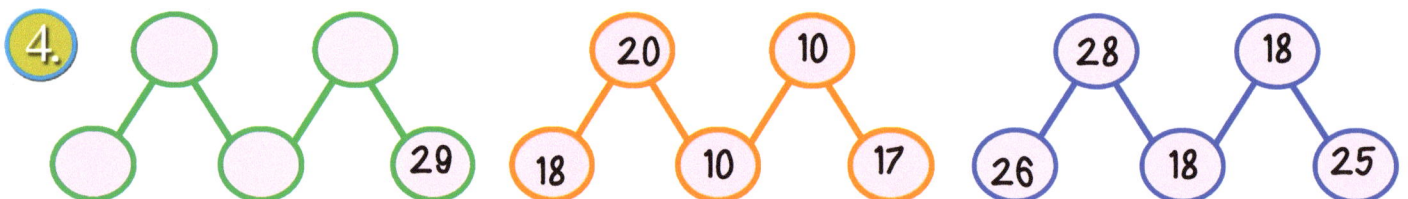

1.

Pattern A (pink): top — 24, 16; bottom — 6, 26, 8

Pattern B (purple): top — 40, 32; bottom — 10, 42, 16

Pattern C (cyan): top — ___, ___; bottom — 3, 14, 2

2.

Pattern A (blue): top — 7, 31; bottom — 21, 28, 30

Pattern B (green): top — 10, 43; bottom — 30, 40, 42

Pattern C (orange): top — ___, 19; bottom — 12, ___, 18

3.

Pattern A (cyan): top — 15, 7; bottom — 3, 14, 1

Pattern B (pink): top — 35, 17; bottom — ___, ___, ___

Pattern C (purple): top — 45, 22; bottom — 9, 44, 16

4.

Pattern A (green): top — ___, ___; bottom — ___, ___, 29

Pattern B (orange): top — 20, 10; bottom — 18, 10, 17

Pattern C (purple): top — 28, 18; bottom — 26, 18, 25

Fact Search

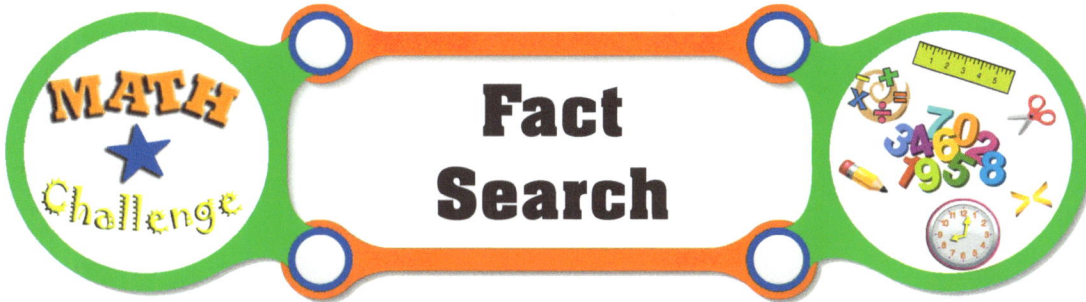

Directions: There are 20 addition, subtraction, division, and multiplication facts hidden in the puzzle below. Can you find all of them? Facts may be horizontal, vertical, or diagonal. Be sure to write each fact you find on the lines.

11	8	15	2	30	2	15
6	11	16	81	11	36	8
15	4	121	9	140	8	7
6	12	31	9	2	9	35
21	15	9	4	70	48	7
6	3	2	108	6	15	5
32	2	24	8	21	3	18
56	3	63	6	24	45	6
8	9	72	5	12	2	24
7	25	10	30	10	3	80

1._____

2._____

3._____

4._____

5._____

6._____

7._____

8._____

9._____

10._____

11._____

12._____

13._____

14._____

15._____

16._____

17._____

18._____

19._____

20._____

Directions: Complete each multiplication problem below by finding the missing digits.

1.

```
  □ 3
x   5
------
3 1 5
```

2.

```
  □ □ 1
x     4
--------
1, 2 8 4
```

3.

```
  8 7
x   □
------
6 9 6
```

4.

```
  2 □
x □ 5
------
1 1 5
2 3 0
------
3 4 5
```

5.

```
  □ 4
x 6 □
--------
1 8 8
5 6 □ 0
--------
5, 8 2 8
```

Class Gardens

The fifth grade classes at an elementary school are allowed to build and maintain a small garden. Use each class garden pictured below to answer the questions.

Class A

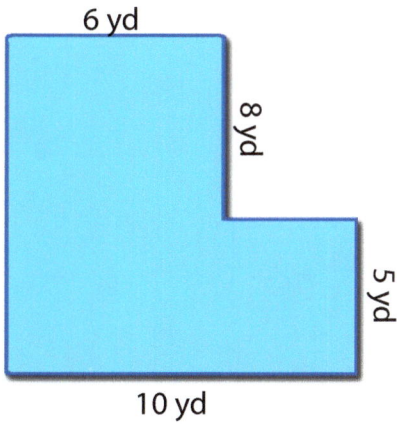

6 yd
8 yd
5 yd
10 yd

Class C

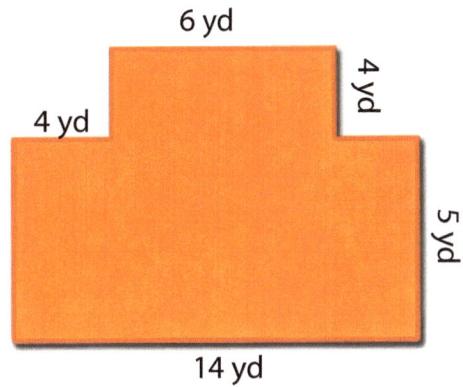

6 yd
4 yd
4 yd
5 yd
14 yd

Class B

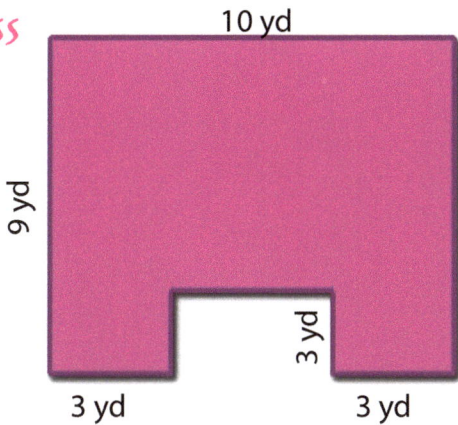

10 yd
9 yd
3 yd
3 yd
3 yd

Class D

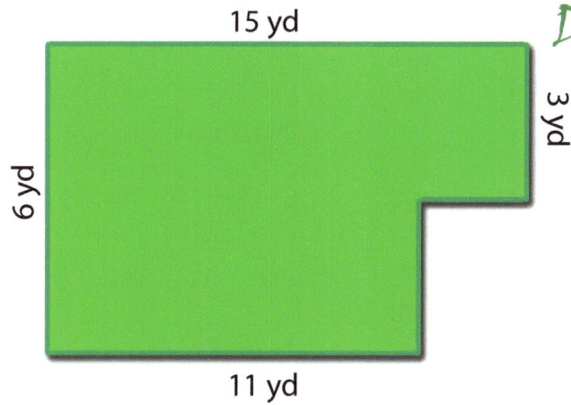

15 yd
3 yd
6 yd
11 yd

1. Which class gardens have the same area?

_____

2. Which class gardens have the same perimeter?

_____

3. Which class used the least amount of wood to build a fence for its garden?

_____

4. Which class has the second largest area available to garden?

_____

MATH Challenge 2

Wow, Super Congratulations!!

_____
(Name)

has completed and mastered the
Math Challenge Level III!

_____
(Signature)

_____
(Date)

Directions: Devise your own pattern to create each puzzle below. Provide two examples and leave one set of popcorn kernels blank for the answer. Challenge a friend to solve your pattern correctly!

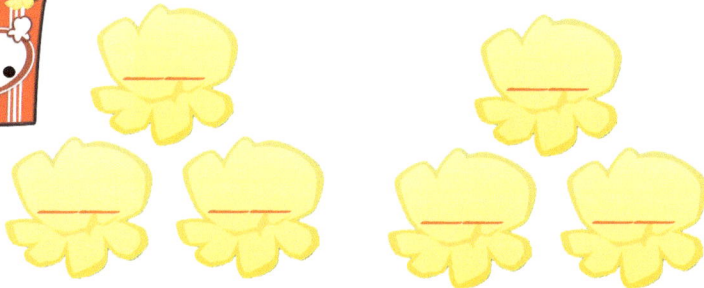

1.

2.

3.

Directions: Use the symbols below to create a "Shapes & Sums" puzzle. Challenge a friend to solve it!

*Tips: The sums for each row and column belong in the shaded boxes. At least one row and one column should be completely filled with shapes and include the sum.

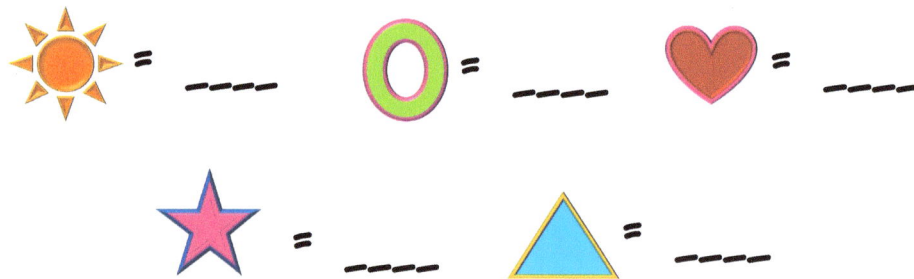

☀ = ____   O = ____   ♥ = ____

★ = ____   △ = ____

Create Your Own Hexagon Pyramids

To create a Hexagon Pyramid, fill in numbers starting from the bottom row. Each block will be added to the block next to it, and the resulting sum will be placed in the block above them.

Directions: Plan the numbers needed to create the puzzle below. Challenge a friend to solve it!

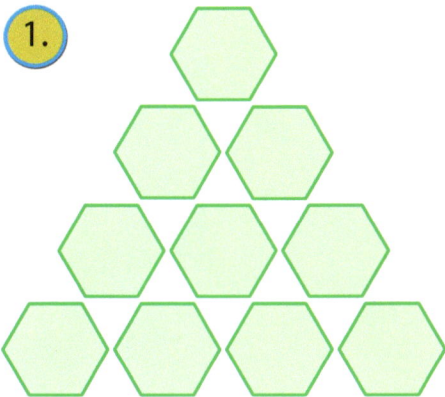

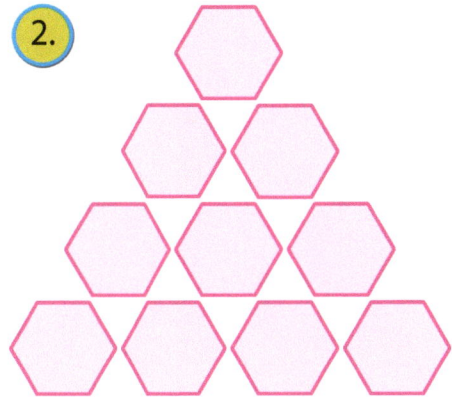

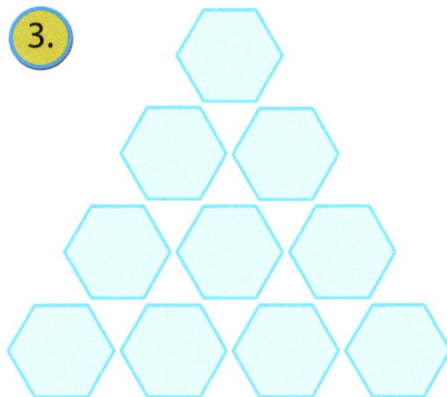

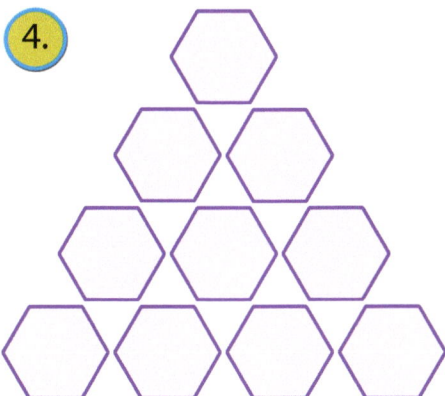

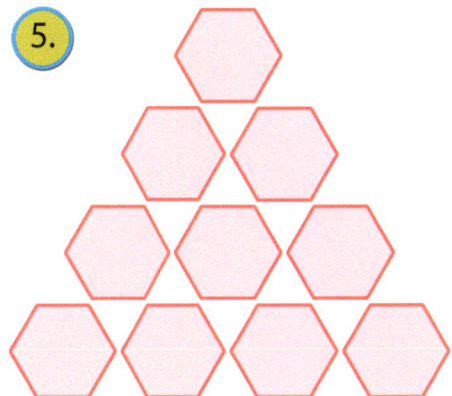

1.

2.

3.

4.

5.

Create Your Own
Rolling Dice

Directions: A group of Math students rolled dice as a part of a class project. Draw dots on each of the "die" below to create a pattern of results. Challenge a friend to predict the next roll in the pattern!

1. ☐☐ , ☐☐ , ☐☐ , ☐☐

2. ☐☐ , ☐☐ , ☐☐ , ☐☐

3. ☐☐ , ☐☐ , ☐☐ , ☐☐

4. ☐☐ , ☐☐ , ☐☐ , ☐☐

5. ☐☐ , ☐☐ , ☐☐ , ☐☐

Answer Key

pg. 4- Popcorn Patterns

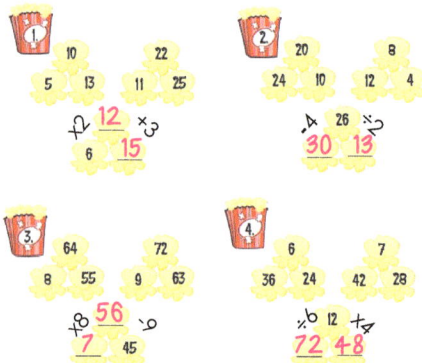

Popcorn 1: 10, 22, 5, 13, 11, 25
+2 12 ×3
6 15

Popcorn 2: 20, 8, 24, 10, 12, 4
-4 26 ÷2
30 13

Popcorn 3: 64, 72, 8, 55, 9, 63
+8 56 -9
7 45

Popcorn 4: 6, 7, 36, 24, 42, 28
÷6 12 +4
72 48

pg. 5- Pattern Word Problems

1.] Day 3 = 6
Day 4 = 11
Day 5 = 18
Day 6 = 27
Day 7 = 38

2.] Level 5 = 405

3.] July = 104
Aug = 112

4.] week 1 = 22
week 4 = 58
week 5 = 70
week 8 = 106

pg. 6- Place Value Puzzle

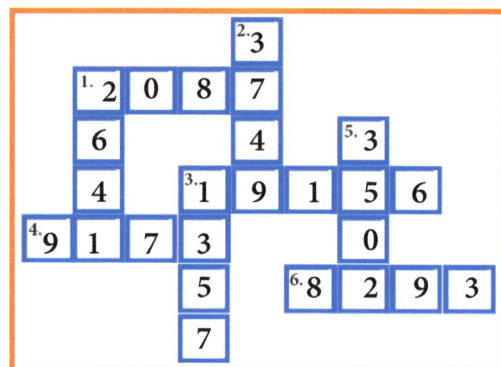

pg. 7- Missing Digits

1. 867 + 121 = 988
2. 612 + 283 = 895
3. 128 + 349 = 477
4. 604 + 276 = 880
5. 135 + 420 = 555
6. 571 + 329 = 900

pg. 8- Hexagon Pyramids

1. 52, 28, 24, 12, 14, 10, 7, 5, 9, 1
2. 83, 37, 46, 10, 27, 19, 5, 15, 12, 7
3. 241, 109, 132, 39, 70, 18, 21, 49, 13
4. 963, 500, 463, 294, 206, 257, 199, 95, 111, 146
5. 684, 515, 169, 387, 128, 41, 299, 88, 40, 1

pg. 9- Pulling Cards
possible answers

1. 8 x 2 + 9 - 5 = 20
2. 9 + 5 + 8 - 2 = 20
3. 4 + 5 + 9 + 2 = 20
4. 7 - 5 + 8 x 2 = 20
5. 9 - 4 + 5 x 2 = 20

pg. 10- Hanging Pattern

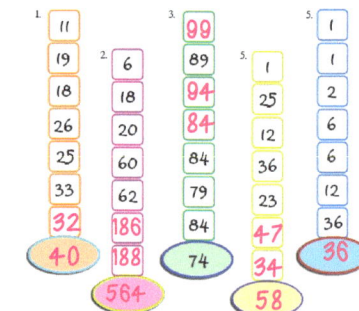

1. 11, 19, 18, 26, 25, 33, 32, 40
2. 6, 18, 20, 60, 62, 186, 188, 564
3. 99, 89, 94, 84, 84, 79, 84, 47, 34, 74
4. 1, 25, 12, 36, 23, 58
5. 1, 2, 6, 6, 12, 36, 36

pg. 11- Symbol Solve

1. ♥ = 0
2. □ = 4
3. △ = 5, ● = 1, □ = 17
4. ♥ = 6, ⬡ = 4, ■ = 10, ▲ = 30

pg. 12- Building Numbers

1. 234
2. 23, 794
3. 97, 423
4. 9, 4793
5. 49, 732

pg. 13- Working Backwards

1. 10
2. 21
3. 100
4. 5
5. 90

pg. 15- Price Tag Riddles

1. $8.42
2. $7.63
3. $57.92
4. $387.41
5. $1, 065.42

Answer Key

MATH ★ Challenge

pg. 16- Cellular Patterns

1. 50; (Jan = 35, Feb = 15)
   The pattern: The difference between 2 months decreases by 20 *(Ex: the difference between Sept/Oct is 100, the difference between Oct/Nov is 80, etc.)*

2. May = $53.75, Sept. = $98.75
   The pattern: Ea. month increases by $11.25

3. Month 7: Ethan = 75   Month 8: Chad = 66
   Month 9: Chad = 30   Month 10: Ethan = 35
   The pattern: Each month alternates with (Ethan ÷ 3 = Chad), and (Ethan x 2 = Chad)

4. 635; (Dec = 639, May = 4)
   The pattern: The increase in sales doubled each month *(Ex: Sales from June to July increased by 10, Sales from July to Aug increased by 20, etc.)*

pg. 17 - Missing Digits 2

pg. 18- Rolling Dice

The sum of the pair of dice is important to find the next roll.

1. Sum: 11
   Pattern: The sum of each roll increases by 2

2. Sum: 4
   Pattern: The sum of each roll decreases by 2

3. Sum: 3
   Pattern: The sum of each roll follows (÷2), then (+2) pattern

4. Sum: 9
   Pattern: The sum of each roll follows (-2), then (+1) pattern

pg. 20- Shapes & Sums

☀ = 6   〇 = 10   ♥ = 12
★ = 5   ▲ = 13

pg. 20- Pulling Cards 2

1. (3 x 5) + 10 = 25
2. 7 + (6 x 3) = 25
3. 8 + 9 + 3 + 5 = 25
4. 6 x 4 + 8 - 7 = 25
5. (9 x 7) - (8 + 5) ÷ 2 = 25
6. 7 + (9 x 2) ÷ (4 - 3) = 25

pg. 21- Symbol Solve 2

1. ☆ = 3
   ⬡ = 15
   ▲ = 60
   ▧ = 4
   〇 = 4

2. ▲ = 30
   ● = 2
   ▢ = 8

3. ♥ = 48
   ⬡ = 24
   ▢ = 120
   ▲ = 8

pg. 22- Who Am I

1. 39
2. 438
3. 153 or 135

pg. 24- PEMDAS Sentences

1. 8 x 3 - 1 = 23
2. (8 - 5) x 2 = 6
3. 4 - (5 - 3) = 2
4. 9 ÷ 3 + 5 = 8
5. 16 = (8 x 2) x 1
7. 12 ÷ 3 x 2 = 2
8. 14 = 2 x (5 + 2)
9. 5 + 3 = 10 ÷ 2
10. 6 - (8 ÷ 2) = 2

pg. 25- Complete the Chart

Jot down the value of each shape:
♥ = 4   ★ = 9
▲ = 6   ☀ = 13

How to solve the chart:

Solve for these in order:
1. Row 3, find value of the heart
2. Column 1, find value of triangle
3. Column 1, find value of star
4. Row 4, find value of sun
5. Find missing values in Column 3
6. Find missing values in Rows 2 & 5
7. Find missing value in Column 2
8. Find missing value in Row 1

MATH ★ Challenge — Answer Key

pg. 26- Rolling Dice 2

Patterns:

1. die 1= (die 2 - die 1)
   die 2: (decreases by 1)

2. die 1: (decreases by 1)
   die 2: (increases by 1)

3. die 1= (alternates +3, -3)
   die 2 = (increases by 1)

4. die 1: (decreases by 1)
   die 2: (alternates -2, +1)

5. die 1= (alternates -2, +1)   die 2 = (alternates +3, -2)

pg. 27- Ladder Climb

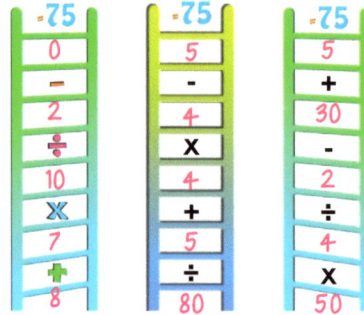

-75	-75	-75
0	5	5
—	-	+
2	4	30
÷	X	-
10	4	2
X	+	÷
7	5	4
+	÷	X
8	80	50

pg. 28- Wave Patterns

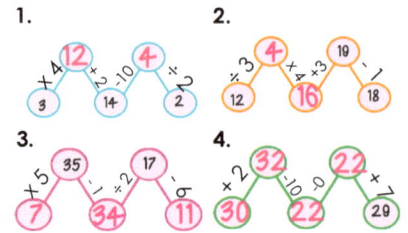

1. ×4: 12, +2: 4; 3, -10: 14, +2: 2
2. ÷3: 4, ×4: 19, +3: 16; -1: 12, 16, 18
3. ×5: 35, -1: 17; 7, +2: 34, -6: 11
4. +2: 32, 22; ÷2: 30, -10: 22, +7: 29

pg. 29- Fact Search

(word search grid)

pg. 30- Missing Digits 3

1. [6]3 × 5 = 315
2. [3][2]1 × 4 = 1,284
3. 8 7 × [8] = 696
4. 2[3] × [1]5: 115, 230 = 345
5. [9]4 × 6[2]: 188, 56[4]0 = 5,828

pg. 31- Class Gardens

1. Class B & Class D
2. Class A & Class C
3. Class D
4. Class C

How to Find Area of Class Gardens:

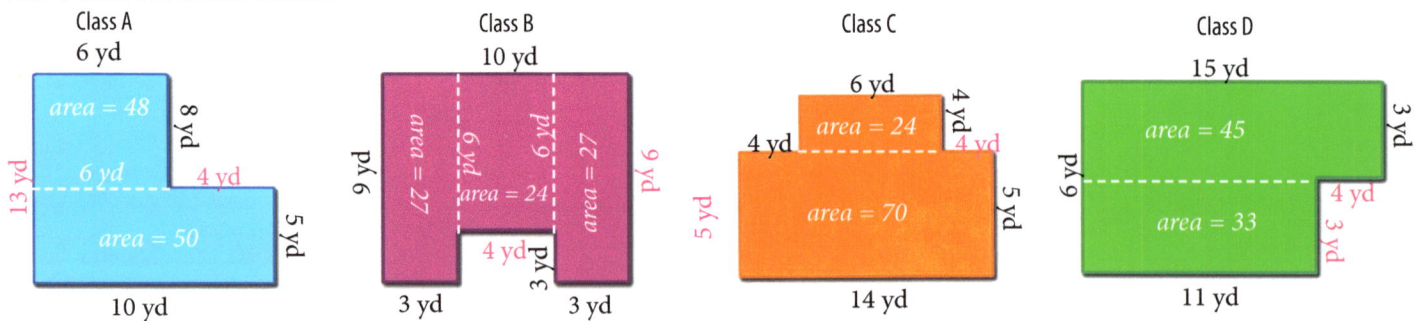

Class A: 6 yd, 8 yd, area = 48; 13 yd, 6 yd, 4 yd; area = 50; 5 yd; 10 yd

Class B: 10 yd, 9 yd, area = 27, area = 24, area = 27; 4 yd, 3 yd, 3 yd, 9 yd

Class C: 6 yd, 4 yd, area = 24; 4 yd, 4 yd, area = 70, 5 yd; 14 yd

Class D: 15 yd, 3 yd, area = 45; 9 yd, area = 33, 4 yd, 3 yd; 11 yd